How [...] de

If you are [...] just a curious
traveler v[...] you may not
realize tha[...] entifying traits.
Often, determining the identity of a rock or mineral is as simple as
trying to scratch it with a pocket knife, or noting how "shiny" it is.
And once you've identified what you've collected, you've taken the
first step toward starting your own collection and becoming a "rock
hound." This book will help you learn to identify 56 of the most
common rocks and minerals in the United States, as well as a few
rarer varieties.

Rocks and minerals are closely related but very different.
Minerals can be thought of as pure substances consisting of a
definite chemical compound. Because of this uniform composition,
minerals **crystallize**, or harden, into definite shapes. **Rocks**, on the
other hand, contain a mixture of different minerals. They form in a
number of ways but always occur as masses of grain-like mineral
particles; rock masses can range from microscopic to quite large.

A crystal
of calcite,
a mineral

A sample
of gabbro,
a rock

How do I identify a rock?

Studying rocks can get very complicated, and identifying rarer rock
types will require you to do lots of research. But some of the sim-
plest and most abundant kinds of rock can be identified by noting
their hardness, color, texture and grain size. A rock's **hardness** can
vary, but knowing whether a pocket knife can scratch a specimen
often helps in identification. One of the most diagnostic traits is
grain size, which is a measure of the size of the mineral grains of
which the rock is composed. Granite, with its large, chunky, easily
visible mineral grains, is an example of a coarse-grained rock. Rhyolite is an example of a very fine-grained rock, with mineral grains so
small that the rock appears uniform in color. Grain size is generally
only used to describe volcanic rocks (which formed when molten
rock solidified), but for the purposes of this guide, it will be used to
describe the texture of all rocks.

Granite, a coarse-
grained rock
(actual size)

Diabase, a medium-
grained rock
(actual size)

Rhyolite, a fine-
grained rock
(actual size)

How do I identify a mineral?

There are thousands of known minerals, and telling one from another may seem like a daunting proposition. But unlike rocks, minerals have very consistent traits that are easy to test and study. Only a handful of minerals are truly considered "common" and easily found by amateurs. To identify these, hardness, color, luster and crystal shape are very helpful traits to note. The **hardness** of a mineral can be determined by scratching it with a tool, such as a pocket knife or your fingernail. When trying to scratch a mineral, it is important to not apply too much pressure, as this can give you a false result. The tool must easily "bite" into the mineral without high pressure. **Luster** is also very helpful, and it describes how "shiny" a mineral is. For example, some minerals are dull, some are glassy, and many others are obviously metallic. But a mineral's **crystal shape** is the most important trait to watch for. When a mineral forms, it takes a definite shape, so a specimen of the same mineral, even from a different part of the world, can share the same identical shape. This often makes a mineral's shape critical when identifying it. The exception is when a specimen is broken or when it formed **massively** (with no crystal shape evident, such as when it formed in a tight space). In these cases, you'll have to rely on other characteristic traits. Note: when geologists use the term "massive," they are rarely talking about physical size.

A well-formed crystal of dull feldspar

A massive sample of metallic chalcopyrite

How to use this guide:

First, determine whether your specimen is a rock or a mineral and go to the appropriate section of this guide. Then compare your rock or mineral to the photos, and see if it matches the description that follows. If you're having a hard time distinguishing between two minerals—between quartz and calcite, for example—read the "How are they different?" text, as this will teach you practical ways to tell them apart. If you can't seem to find your specimen in the book, you may have something more uncommon; see the note at the back of this book.

Is there anything that I can't collect?

Collecting anything in national parks, on Native American reservations, or in some state and national forests is illegal. In addition, you'll be trespassing if you collect on privately owned property, so always be aware of where you are collecting.

Most states have rules that prohibit collection of vertebrate fossils (fossils of animals with a backbone, such as fish, reptiles and dinosaurs) due to their scientific importance. It is also illegal to collect Native American artifacts, which should instead be reported to authorities. Familiarize yourself with your state's collecting laws before heading into the field.

Coarse-grained Rocks

Granite
A very common rock, granite underlies most of North America

- Hard; not typically scratched with a pocket knife
- Dense and coarse-grained with mottled coloration; gray, yellow, brown, pink and black
- Contains large, angular, embedded crystals, some of which appear glassy or nearly metallic
- Found nearly anywhere, but especially in mountainous regions

How are they different?
Granite's more common, and its light-colored grains are larger and more prevalent

Diorite
A fairly common darker volcanic rock that resembles granite

- Hard; not typically scratched with a pocket knife
- Dense and coarse-grained with mottled coloration; black, gray and white
- Contains fewer light-colored mineral grains than granite
- Best found in mountainous regions, especially in the West and Northwest

How are they different?
Diorite is coarser grained and contains more light-colored grains than diabase

Diabase
An uncommon dark, dense, medium-grained volcanic rock

- Hard; not typically scratched with a pocket knife
- Dark in color with medium-sized mineral grains, some of which may appear glassy
- Contains small embedded light-colored mineral grains, sometimes vaguely circular in shape
- More common in mountainous regions, such as the West and Northeast

How are they different?
Gabbro is much coarser grained and contains much larger crystals than diabase

Gabbro
An uncommon dark, greenish, coarse-grained volcanic rock

- Hard; not typically scratched with a pocket knife
- Very dark color, typically nearly black and often with a greenish tint
- Contains large embedded rectangular crystals that are very reflective in certain directions
- Mountainous regions, such as in the West; also found in the Lake Superior region

Coarse-grained Rocks

Porphyry
Uncommon rocks containing large, very conspicuous crystals

- Fairly hard; may be scratched with a pocket knife
- Can resemble granite, basalt or rhyolite but has well-formed angular crystals embedded within it
- Crystals are often lighter colored and are conspicuous, appearing "out of place"
- Most common in mountainous regions, such as the Northwest and Northeast

How are they different?
Porphyry contains crystals of a fairly uniform size and shape embedded in hard rock

Conglomerate and Breccia
Common rocks consisting of smaller rocks stuck together

- Both rocks vary greatly in hardness
- Conglomerate is a rock consisting of small rounded stones cemented together
- Breccia is similar but consists of broken, angular rock fragments cemented together
- Abundant in the Midwest; also often found in foothills and river valleys

How are they different?
Conglomerate and breccia are normally much coarser grained and harder than tuff

Tuff
A common rock composed of compacted volcanic ash

- Soft; often easily scratched with a pocket knife
- Light gray to brown in color; finer grained, with a very gritty texture and occasional darker spots
- Frequently contains small fragments of dark volcanic glass (obsidian) embedded within
- Very common in the western U.S., especially in the Southwest

How are they different?
Tuff is harder and does not expand or become sticky when wet, as clay does

Clay
Very common, soft, crumbly, fine-grained sediment

- Very soft; easily scratched with your fingernail
- Extremely fine-grained; varies in color, but typically gray to brown, sometimes with layering
- Crumbly when dry, but malleable and sticky when wet; often contains embedded pebbles
- Very widespread; abundant near any body of water, especially in river valleys

Fine-grained Rocks

Rhyolite
A very common hard, fine-grained volcanic rock

- Very hard; can't be scratched with a pocket knife
- Fine-grained and light-colored, often in shades of gray, brown or reddish
- May have parallel banding, veins of minerals, and/or many gas bubbles
- Very widespread; found in many states, especially in the West

How are they different?
Basalt is always darker than rhyolite; basalt typically doesn't show any banding

Basalt
A common, dark and dense volcanic rock

- Fairly hard; may be scratched with a pocket knife with some effort
- Very fine-grained and always dark in color, from gray to greenish black
- Often contains bubble-shaped cavities
- Widespread; common in the Southwest, Northwest and Northeast

How are they different?
Chert is far harder, typically lighter in color, and often has a waxy appearance

Chert
A very common and extremely hard, waxy rock

- Very hard; can't be scratched with a pocket knife
- Typically light gray to brown or black; always opaque, and sometimes with colored layers
- Breaks into very sharp-edged pieces; worn pieces appear smooth and waxy, as if polished
- Chert is abundant virtually anywhere, especially in river valleys

How are they different?
Obsidian is far more glass-like, and chert is more opaque and more common

Obsidian
An uncommon, black, translucent, shiny volcanic glass

- Hard; not typically scratched with a pocket knife
- Very brittle, glassy and dark, typically black or dark brown in color; translucent when thin
- Breaks in circular shapes when struck; broken fragments have extremely sharp edges
- Most common in western states, particularly in mountainous areas

Fine-grained Rocks

Schist
A common, dense and highly layered metamorphic rock

- Fairly hard; may be scratched with a pocket knife with some effort
- Dense and highly layered; typically gray and may contain countless tiny shiny flecks
- Can contain large, very hard minerals embedded within, such as garnet
- Common in hilly or mountainous areas

How are they different?
Schist has much more compact layers and is often "glittery" with shiny grains

Gneiss
Gneiss (pronounced "nice") is a common metamorphic rock

- Fairly hard; may be scratched with a pocket knife with some effort
- Vaguely layered with bands of varying coloration; it has both coarse and fine mineral grains
- May resemble other rocks, particularly granite, but in a layered or banded form
- Best found in mountainous or hilly areas

How are they different?
Quartzite is glassier and more uniform in texture and hardness

Quartzite
A common, extremely hard and dense metamorphic rock

- Very hard; can't be scratched with a pocket knife
- Generally light in color, often with faint layering, and typically translucent, especially at the edges
- Grainy and glassy when freshly broken; smooth when weathered
- Common everywhere, especially in the Northwest

How are they different?
Marble is much softer and typically does not exhibit layering; marble is rarer

Marble
An uncommon metamorphic rock formed from limestone

- Soft; easily scratched with a pocket knife
- Exhibits light coloration, sometimes purely white; virtually always opaque
- Larger, glassy grains are sometimes present and will "fizz" in vinegar
- Uncommon; found in hilly or mountainous regions

Fine-grained Rocks

Shale
A common, soft, layered rock found in low-lying areas

- Soft; can be scratched with a pocket knife
- Highly layered; layers are easy to split apart with a knife blade
- Fine-grained and typically brown in color; may have fossils embedded between layers
- Common in the Midwest, the East and in any flat regions

How are they different?
Shale and mudstone are very similar, but shale is highly layered; mudstone is not

Mudstone
A common, very soft sedimentary rock with no layering

- Very soft, easily scratched with a pocket knife
- Resembles shale in many ways, but is not layered
- Extremely fine-grained with an even coloration; typically gray to tan or brown
- Common in the Midwest, the Northwest, and any flatter regions

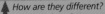

How are they different?
Limestone is harder, more common and often lighter in color

Limestone
A very common, light-colored sedimentary rock

- Soft; can be scratched with a pocket knife
- Light-colored, sometimes with layers of varying coloration; has a gritty texture
- Will "fizz" in vinegar and sometimes contains crystal-lined cavities
- Very common; found almost anywhere in the U.S., especially the Midwest

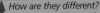

How are they different?
Sandstone is coarser than limestone and generally doesn't "fizz" in vinegar

Sandstone
One of the most common sedimentary rocks

- Soft; can be scratched with a pocket knife
- Composed of cemented sand, so it has a coarse, gritty, sandy texture
- Typically light-colored, but often with darker colored layers or bands
- Widespread; abundant in low-lying regions such as the Midwest and Southeast

Dull to Glassy Minerals

Quartz
The most common mineral; found in every environment

- Very hard; can't be scratched with a pocket knife
- Light-colored (typically colorless to white), glassy and translucent
- Crystals have six sides and a pointed tip; masses are more common, often found loose as pebbles or embedded in rocks
- Quartz can be found literally anywhere

 How are they different?
Quartz is much, much harder than calcite; calcite breaks more readily

Calcite
A very common mineral, particularly in sedimentary rocks

- Soft; easily scratched with a pocket knife
- Light-colored (typically colorless to white), translucent and often glassy
- Forms crystals shaped as steep six-sided points or blocky "leaning" cubes; will "fizz" in vinegar
- Very widespread; can be found in every state, especially as masses or veins in rocks

 How are they different?
Dolomite is less common than calcite, and its crystals tend to be more opaque

Dolomite
A common mineral typically found in limestone

- Soft; easily scratched with a pocket knife
- Light-colored, with a pearly luster; sometimes translucent, but more often opaque
- Crystals form as blocky "leaning" cubes but often exhibit curved or rounded sides
- Prevalent in low-lying, sedimentary regions, such as the Midwest and Southwest

 How are they different?
Gypsum is far softer than dolomite and will dissolve very slowly in warm water

Gypsum
A very common, extremely soft mineral

- Very soft; easily scratched with your fingernail
- Light-colored, typically white to tan; may appear glassy, fibrous or grainy
- Forms elongated needle-like crystals, or as masses in rock; will slowly dissolve in warm water
- Very common in flatter areas, such as the Midwest and in the deserts of the Southwest

Dull to Glassy Minerals

Barite
A soft mineral so dense that even small pieces feel heavy

- Soft; easily scratched with a pocket knife
- Light-colored, translucent and often glassy; very brittle
- Often found as thin blades or fat, wedge-shaped points that feel very heavy for their size
- Found in mountainous regions across the U.S.

↕ *How are they different?*
A piece of barite has more "heft" than any similarly sized piece of fluorite

Fluorite
A fairly common mineral found in a rainbow of colors

- Soft; easily scratched with a pocket knife
- Varies in color; often white, green, purple or blue; typically glassy and translucent
- Crystals are found as cubes or octahedrons (which look like two pyramids put bottom-to-bottom)
- Fluorite is widespread and is found in most states, often in limestone

 How are they different?
Olivine is far harder and generally yellow-green in color; olivine does not form cubes

Olivine
A hard mineral most common as grains in dark rocks

- Hard; can't be scratched with a pocket knife
- Typically greenish yellow in color and very glassy and translucent
- Most often found as small embedded grains in dark rocks like gabbro; also found loose in sand
- Widespread; abundant in mountainous areas, particularly the Northwest and Rockies

↕ *How are they different?*
Olivine is more common and generally more richly colored than prehnite

Prehnite
An uncommon but highly collectible mineral

- Hard; can't be scratched with a pocket knife
- Typically gray-green to apple-green in color, often glassy and generally always translucent
- Crystals are wedge-like, often with rounded edges; they can also occur as ball-like clusters
- Not common, but found around Lake Superior and in the Northeast

Dull to Glassy Minerals

Epidote
An uncommon and highly collectible mineral

- Very hard; can't be scratched with a pocket knife
- Virtually always a distinctive yellow-green or olive-green color; somewhat translucent
- Crystals are elongated with grooved sides; also forms as crusts or veins in rock
- Epidote is found across the U.S., typically in mountainous regions

 How are they different?
Epidote is much harder than the serpentines; serpentines are not found as crystals

Serpentine group
A fairly common group of soft, "greasy" minerals

- Soft; easily scratched with a pocket knife
- Serpentines are green to yellow or brown in color and opaque
- Forms as irregular masses that feel "greasy"; rarer serpentines appear as masses of fibers
- Most abundant near mountainous regions, as well as near the West Coast

 How are they different?
Serpentine minerals are harder than talc; talc can be scratched by your fingernail

Talc
Talc is the softest mineral, though it is not very common

- Extremely soft; can be scratched with your fingernail with little effort
- Light-colored, opaque and white to pale green
- Forms as irregular flaky masses that are chalky and feel like soap
- Talc is most abundant in hilly or mountainous regions where schist is prevalent

 How are they different?
Talc is lighter in color and forms as large masses rather than as tiny crystals

Chlorite group
Very common, very soft minerals found coating rock cavities

- Very soft; easily scratched with your fingernail
- Dark-colored in shades of green or brown, appearing "greasy" in luster
- Forms as thin coatings or tiny six-sided crystals within cavities in rock, particularly in basalt
- Very common in the West, Upper Midwest and Northeast

Dull to Glassy Minerals

Azurite
An uncommon, vividly colored copper-bearing mineral

- Soft; easily scratched with a pocket knife
- Always vivid blue in color, often very dark colored and brightly lustrous; glassy
- Forms as blocky, pointed crystals when well formed; also occurs as gritty coatings or as balls
- Azurite is most prominent in the Southwest

How are they different?
When compared with azurite, turquoise is harder, rarer and not as dark blue

Turquoise
An uncommon and very popular copper-bearing mineral

- Fairly hard; may be scratched with a pocket knife with some effort
- Light blue and opaque, typically dull
- Generally forms as irregular masses or veins within cavities in rock
- Turquoise is fairly rare and is virtually only found in dry areas, such as the deserts of the Southwest

How are they different?
Chrysocolla is far more common and widespread and is also much softer

Chrysocolla
A soft, common mineral that forms when copper weathers

- Soft; easily scratched with a pocket knife
- Pale blue to bluish green in color, often growing on copper or other copper-bearing minerals
- Typically forms as a thin powdery crust or coating; also sometimes occurs as veins in rock
- Common in the Southwest, as well as in northern Michigan

How are they different?
Malachite is harder and exhibits a fibrous cross section; chrysocolla is more crumbly

Malachite
An attractive and abundant copper mineral

- Soft; can be scratched with a pocket knife
- Vivid green in color; sometimes exhibits banding
- Forms as lumpy, rounded, fibrous masses or crusts on or near copper; more rarely forms as hair-like crystals
- Fairly abundant in the Southwest, as well as in northern Michigan

Dull to Glassy Minerals

Feldspar group
The most abundant group of minerals, found in many rocks

- Hard; can't be scratched with a pocket knife
- Light-colored, often white, gray or pink (rarely colorless); typically opaque or slightly translucent
- Most commonly found embedded in rocks like granite as blocky, angular grains or masses
- Feldspars are extremely common and can literally be found anywhere in the U.S.

How are they different?
Feldspars are far more common than tourmalines, which tend to be darker

Tourmaline group
A fairly uncommon group of attractive minerals

- Very hard; can't be scratched with a pocket knife
- Most tourmalines are black, glassy and opaque; a rarer variety is pink and translucent
- Crystals are slender and elongated with striated (grooved) sides and a triangular cross section
- Most common in mountainous regions in the West and Northeast

How are they different?
Tourmaline crystals are elongated, while garnet crystals tend to be ball-like

Garnet group
A common family of very hard, colorful minerals

- Very hard; can't be scratched with a pocket knife
- Garnets are most commonly red to brown, translucent and glassy
- Crystals appear as faceted "balls," often embedded in schist or granite
- Crystals are common in mountainous regions, as well as in river gravel

How are they different?
Garnets are much harder and more common than sphalerite, which is often darker

Sphalerite
A fairly common ore of zinc that forms attractive crystals

- Soft; easy to scratch with a pocket knife
- Typically darkly colored and very brightly lustrous; ranges from dark red or deep yellow to nearly black
- Crystals are complex shapes, often with triangular features; masses may have a "velvety" sheen
- Widespread; the Midwest and Rocky Mountain states have produced fine specimens

Dull to Glassy Minerals

Opal
A common glassy material that does not form crystals

- Fairly hard; not typically scratched with a pocket knife
- Light-colored, often white and opaque; exhibits many glass-like traits
- Does not form crystals; typically found as veins or pockets in volcanic rocks, such as rhyolite
- Most abundant in the West

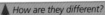

How are they different?
Opal is slightly softer and is often lighter in color; jasper is not as lustrous

Jasper
A very common and colorful variety of chert

- Hard; can't be scratched by a pocket knife
- Found in any color, though typically brown to red and often mottled; opaque unless very thin
- Forms as veins or pockets in rock; often found as loose, waxy rounded pebbles
- Very common across the U.S., especially on coastal or inland beaches; also found in river valleys

How are they different?
Jasper is always more opaque than chalcedony under bright light

Chalcedony
A very common variety of dense, compact quartz

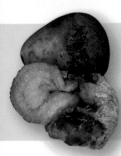

- Hard; can't be scratched by a pocket knife
- Color can vary greatly, though brown to red is common; specimens are often mottled
- Forms as pockets or veins in rocks; often found as loose, waxy, translucent rounded pebbles
- Very common across the U.S., especially on beaches or along rivers; less common in the Southeast

How are they different?
Chalcedony's color is often mottled, while agates are organized in ring-like bands

Agates and Thunder Eggs
Uncommon banded gemstones of mysterious formation

- Hard; can't be scratched by a pocket knife
- Color varies greatly; multicolored in concentric rings; typically found in red, brown or gray
- Found as waxy colorful pebbles in rivers or gravel; "thunder eggs" have a rocky exterior
- Most common west of the Mississippi River and in the Lake Superior region

Metallic Minerals

Hematite
The most common iron-bearing mineral

- Fairly hard; may be scratched with a pocket knife with some effort
- Metallic-gray in color, turning rusty red when weathered or powdered
- Takes a variety of forms, occurring as crusts on rock, lumpy rounded masses, or thin blade-like crystals
- Very common, found virtually anywhere

How are they different?
Hematite turns reddish when crushed and powdered; goethite turns orange-brown

Goethite
Pronounced "ger-tite," this is a common iron mineral

- Fairly hard; can barely be scratched with a knife
- Metallic black in color, but often coated in a dusty orange or yellow-brown material
- Forms as lumpy rounded masses with a fibrous cross section or as thin rust-brown crusts
- Common and found nearly anywhere, primarily as a dusty rust-like coating on rocks

How are they different?
Goethite is not magnetic and will not attract a magnet; magnetite will

Magnetite
A common iron mineral that is strongly magnetic

- Fairly hard; not easily scratched by a pocket knife
- Black and metallic; will strongly attract a magnet
- Takes the form of octahedrons (crystals that resemble two pyramids placed bottom-to-bottom) or as irregular shiny-to-dull masses
- Widespread; abundant anywhere as grains in sand and gravel; can easily be found with a magnet

How are they different?
Magnetite is magnetic; any piece of galena is very heavy for its size

Galena
The primary ore of lead, galena is very heavy and common

- Very soft; can be easily scratched with a pocket knife
- Gray to bluish gray in color and highly metallic, particularly on a freshly broken surface
- Breaks into perfect cubes; even a small piece will be very heavy for its size
- Very abundant in the central U.S.

Metallic Minerals

Copper
A fairly common native element

- Very soft; easy to scratch with a pocket knife
- Always reddish metallic in color unless weathered; weathered specimens turn green or black
- Highly malleable (easily bent) and not brittle
- Not common everywhere but abundant in the Southwest and in northern Michigan

How are they different?
Pyrite is harder, brittle, brassy in color and much more common

Pyrite
"Fool's gold" is one of the most common metallic minerals

- Very hard; cannot be scratched by a pocket knife
- Brassy yellow in color and very lustrous; sometimes coated in a dull brown material
- Develops as cube-shaped crystals that are often embedded in rocks; very brittle
- Quite common, found in most states; especially common in mountainous regions

How are they different?
Chalcopyrite is softer than pyrite and does not form cubic crystals

Chalcopyrite
A common copper-bearing metallic mineral

- Soft; easily scratched with a pocket knife
- Brass-like in color, often with an orange tint or a bluish multicolored surface coating
- Most often found as brittle, brightly metallic veins embedded in rocks or quartz
- Fairly common, especially in mountainous regions
- Pronounced *Cal-ke-pye-rite*

How are they different?
Mica minerals are much softer, form as flaky crystals and are not actually metallic

Mica group
A very common constituent of rocks

- Very soft; easily scratched with your fingernail
- Often dark-colored and are typically so shiny they almost look metallic
- Crystals form as flexible paper-thin flakes that grow in stacks; typically occur within rocks such as granite
- Micas are found everywhere, most often embedded in rocks as tiny flecks of "glitter"

Geodes and Fossils

Geodes
Uncommon rock formations that are round and hollow

- Often found as unusually round rocks
- When broken open, a geode's interior is hollow and often lined with many tiny crystals
- Interior crystals are often calcite or quartz or, less commonly, agate
- Uncommon, but most abundant in the Midwest and West

Animal fossils
Preserved remains of ancient animals

- Traces of animals, especially clams, snails, teeth or coral, embedded in rock
- Most fossils are found in layers of shale or limestone
- Fossils often occur in the same colors as the rock they are embedded within
- Uncommon, but easiest to find in the West

> *How are they different?*
> Animal fossils resemble the hard parts of modern animals

Plant fossils
Preserved remains of ancient plants within rock

- Impressions of plants, especially leaves, twigs, and ferns, embedded in rock
- Most plant fossils are found in between layers of shale
- Plant fossils are typically colored nearly identically to their host rock
- Rare; found in the Midwest and the East

> *How are they different?*
> Petrified wood is often very hard and not typically found within shale

Petrified wood
A fairly common type of fossil that looks exactly like wood

- Rock or jasper with the appearance of living wood
- Often exhibits wood-like features, such as wood grain, bark and tree limbs
- Can be brightly colored, especially when it consists of jasper; jasper specimens are very hard
- Found in almost every state but most abundant in the Southwest

Commonly Misidentified Minerals

Gold
Gold is a very rare precious metal, and many seek it

- Gold is very soft and easily scratched with a knife; it is also malleable (bendable)
- Gold is always brightly metallic yellow
- Pyrite and chalcopyrite share a similar color, but both are far harder and brittle, not malleable
- Most gold is found as minuscule grains in rivers, especially in Alaska and California

Silver
Many silver-colored minerals are mistaken for this rare metal

- Silver is very soft, easily scratched with a knife, and is malleable (bendable)
- Silver darkens to dull gray or black, but a scratch will always reveal its silver color underneath
- Most other silver-colored minerals are very brittle and will break; silver bends rather than breaks
- Most silver is found as thin veins or flecks in rock

Meteorites
Most suspected "meteorites" are actually magnetite

- Meteorites can be found anywhere but are very rare
- Metal-rich meteorites are often magnetic, silver-gray, and have a rough, dark rusty outer crust
- Magnetite, due to its magnetism, is often mistaken for metallic meteorites
- Always assume a specimen is magnetite until verified by a professional!

Slag
A waste byproduct of ore processing and industry

- Slag is often dark, glassy to metallic in appearance, and a piece can be heavy for its size
- Colorful slag glass is often mistaken for other rarer minerals, especially opal and obsidian
- Slag usually has many round bubbles trapped within it (this doesn't occur in minerals); some slag appears "melted" and looks like it froze in place

Final Notes

I can't find my specimen in the book. Does that mean it's rare?
Not likely. There are many rocks and minerals that are considered "common"; the types represented here are simply among the most abundant. In fact, perhaps as much as 90 percent of what a beginning collector will easily find is presented in this guide. So while you may have found something a bit more uncommon than most rocks or minerals, it is more likely that weathering or staining has altered the appearance or hardness of your common specimen and it no longer appears quite as it should.

The reason certain minerals are so much easier to find than others is largely because the usual places where casual collectors find rocks and minerals—the beach, a mountainside, a rocky road, a riverbed, etc.—are very damaging to specimens. Only rocks and minerals that are very hard will survive these conditions long enough to be found. Alternatively, specimens of soft rocks and minerals can be found in such areas if they are abundant enough to be replenished as quickly as they are worn away. In the case of minerals like quartz, both situations may be true. It is hard enough to survive a good deal of weathering and common enough that other quartz specimens are almost always nearby.

Certain minerals were not included in this book because, although common, they are typically only seen as grains embedded within rocks rather than as loose and collectible specimens. But if you still think you've found something rare, then your next step is further research. Use only the information available to you, such as the specimen's hardness, crystal shape, and where you found it, and don't let assumptions or wishful thinking cloud the identification process. (If you're not careful, it's not hard to "convince" yourself that you've found a rare mineral, even though you really haven't.)

Where do I look for rocks and minerals?
Throughout the United States, the best places to look for rocks and minerals are wild, natural places where rocks are exposed. Therefore, beaches and rivers are popular places, as are hills, mountains and deserts. Man-made exposures of rock and gravel are lucrative as well; examples include gravel pits, quarries, mining sites and dirt roads, but beware of private and protected property.

What equipment is helpful to bring?
If you're planning on walking along a beach you won't need much more than a bucket or a backpack to carry your finds, though a camera and a magnifying lens are useful. If you're planning to dig and break rock, however, you'll need thick, sturdy leather gloves, a rock hammer (not a nail hammer), and eye protection, as well as digging implements. Tissues or paper towels are great to protect any fragile specimens. And always stay safe: remember to bring water, a compass or GPS and a cell phone along with you.